This Drone Flight Log Book
Belongs to

_____

_____

Dedication

This Drone Log Book is dedicated to all the drone enthusiasts out there who want to record all their flight information and document their findings in the process.

You are my inspiration for producing books and I'm honored to be a part of keeping all of your drone notes and records organized.

This journal notebook will help you record the details of your drone flights.

Thoughtfully put together with these sections to record: Date, Drone Model Name, Location, Time, Minutes in Flight, Battery, Weather, Temperature, Wind Speed & Direction, Visual Observer, Operator, Flight Mission & Notes.

How to Use this Book

The purpose of this book is to keep all of your Drone notes all in one place. It will help keep you organized.

This Drone Log Book will allow you to accurately document every detail about your drone flights and missions.

Here are examples of the prompts for you to fill in and write about your experience in this book:

1. Date
2. Drone Model Name
3. Location
4. Time - Write the time.
5. Minutes of Flight - Record how long your drone was in flight.
6. Battery - Log the battery used and how long it lasted.
7. Weather - Write what the weather conditions were like.
8. Temperature - Record the temperature.
9. Wind Speed - Log the speed of the wind.
10. Wind Direction - Write which direction the wind is coming from.
11. Visual Observer - Record your crew of visual observers.
12. Operator - Log who controlled the drone.
13. Flight Mission - Write your mission, whether for fun or for a special event.
14. Notes - Document any other additional important information you want such as technical, system maintenance, your flight map, faa, uav, uas regulations, other flights planned, training, etc.

Drone Flight Log Book

Date:

Drone Model Name

Location (from-to)	Time
Minutes of flight	Battery

Weather	Temperature	Wind speed	Wind Direction

Crew	
Visual Observer	Operator

Flight Mission	Notes

Drone Flight Log Book

Date:	

Drone Model Name

Location (from - to)	Time
Minutes of flight	Battery

Weather	Temperature	Wind speed	Wind Direction

Crew	
Visiual Observer	Operator

Flight Mission	Notes

Drone Flight Log Book

Date:

Drone Model Name

Location (from-to)	Time
Minutes of flight	Battery

Weather	Temperature	Wind speed	Wind Direction

Crew	
Visual Observer	Operator

Flight Mission	Notes

Drone Flight Log Book

Date:

Drone Model Name

Location (from - to)	Time
Minutes of flight	Battery

Weather	Temperature	Wind speed	Wind Direction

Crew	
Visiual Observer	Operator

Flight Mission | **Notes**

Drone Flight Log Book

Date:

Drone Model Name

Location (from-to)	Time
Minutes of flight	**Battery**

Weather	Temperature	Wind speed	Wind Direction

Crew	
Visual Observer	Operator
Flight Mission	Notes

Drone Flight Log Book

Date:

Drone Model Name

Location (from - to)	Time
Minutes of flight	Battery

Weather	Temperature	Wind speed	Wind Direction

Crew	
Visiual Observer	Operator
Flight Mission	Notes

Drone Flight Log Book

Date:

Drone Model Name

Location (from-to)	Time

Minutes of flight	Battery

Weather	Temperature	Wind speed	Wind Direction

Crew	
Visual Observer	Operator
Flight Mission	**Notes**

Drone Flight Log Book

Date:

Drone Model Name

Location (from - to)	Time

Minutes of flight	Battery

Weather	Temperature	Wind speed	Wind Direction

Crew	
Visiual Observer	Operator

Flight Mission	Notes

Drone Flight Log Book

Date:

Drone Model Name

Location (from-to)	Time
Minutes of flight	Battery

Weather	Temperature	Wind speed	Wind Direction

Crew	
Visual Observer	Operator
Flight Mission	Notes

Drone Flight Log Book

Date:

Drone Model Name

Location (from - to)	Time
Minutes of flight	Battery

Weather	Temperature	Wind speed	Wind Direction

Crew	
Visiual Observer	Operator

| Flight Mission | Notes |

Drone Flight Log Book

Date:

Drone Model Name

Location (from-to)	Time
Minutes of flight	Battery

Weather	Temperature	Wind speed	Wind Direction

Crew	
Visual Observer	Operator

Flight Mission	Notes

Drone Flight Log Book

Date:

Drone Model Name

Location (from - to)	Time
Minutes of flight	Battery

Weather	Temperature	Wind speed	Wind Direction

Crew	
Visiual Observer	Operator

Flight Mission	Notes

Drone Flight Log Book

Date:

Drone Model Name

Location (from-to)	Time
Minutes of flight	Battery

Weather	Temperature	Wind speed	Wind Direction

Crew	
Visual Observer	Operator

Flight Mission	Notes

Drone Flight Log Book

Date:

Drone Model Name

Location (from - to)	Time
Minutes of flight	**Battery**

Weather	Temperature	Wind speed	Wind Direction

Crew	
Visiual Observer	Operator
Flight Mission	**Notes**

Drone Flight Log Book

Date:

Drone Model Name

Location (from-to)	Time
Minutes of flight	Battery

Weather	Temperature	Wind speed	Wind Direction

Crew	
Visual Observer	Operator
Flight Mission	Notes

Drone Flight Log Book

Date:

Drone Model Name

Location (from - to)	Time

Minutes of flight	Battery

Weather	Temperature	Wind speed	Wind Direction

Crew	
Visiual Observer	Operator

Flight Mission	Notes

Drone Flight Log Book

Date:

Drone Model Name

Location (from-to)	Time

Minutes of flight	Battery

Weather	Temperature	Wind speed	Wind Direction

Crew	
Visual Observer	Operator
Flight Mission	**Notes**

Drone Flight Log Book

Date:

Drone Model Name

Location (from - to)	Time

Minutes of flight	Battery

Weather	Temperature	Wind speed	Wind Direction

Crew	
Visiual Observer	Operator

Flight Mission	Notes

Drone Flight Log Book

Date:

Drone Model Name

Location (from-to)	Time
Minutes of flight	Battery

Weather	Temperature	Wind speed	Wind Direction

Crew	
Visual Observer	Operator
Flight Mission	Notes

Drone Flight Log Book

Date:

Drone Model Name

Location (from - to)	Time
Minutes of flight	Battery

Weather	Temperature	Wind speed	Wind Direction

Crew	
Visiual Observer	Operator

Flight Mission	Notes

Drone Flight Log Book

Date:

Drone Model Name

Location (from-to)	Time
Minutes of flight	Battery

Weather	Temperature	Wind speed	Wind Direction

Crew	
Visual Observer	Operator
Flight Mission	Notes

Drone Flight Log Book

Date:

Drone Model Name

Location (from - to)	Time
Minutes of flight	Battery

Weather	Temperature	Wind speed	Wind Direction

Crew

Visiual Observer	Operator

Flight Mission | Notes

Drone Flight Log Book

Date:

Drone Model Name

Location (from-to)	Time
Minutes of flight	Battery

Weather	Temperature	Wind speed	Wind Direction

Crew	
Visual Observer	Operator
Flight Mission	Notes

Drone Flight Log Book

Date:

Drone Model Name

Location (from - to)	Time
Minutes of flight	Battery

Weather	Temperature	Wind speed	Wind Direction

Crew	
Visiual Observer	Operator

Flight Mission | **Notes**

Drone Flight Log Book

Date:

Drone Model Name

Location (from-to)	Time
Minutes of flight	**Battery**

Weather	Temperature	Wind speed	Wind Direction

Crew	
Visual Observer	Operator
Flight Mission	Notes

Drone Flight Log Book

Date:

Drone Model Name

Location (from - to)	Time

Minutes of flight	Battery

Weather	Temperature	Wind speed	Wind Direction

Crew	
Visiual Observer	Operator

Flight Mission	Notes

Drone Flight Log Book

Date:

Drone Model Name

Location (from-to)	Time
Minutes of flight	Battery

Weather	Temperature	Wind speed	Wind Direction

Crew	
Visual Observer	Operator
Flight Mission	Notes

Drone Flight Log Book

Date:

Drone Model Name

Location (from - to)	Time
Minutes of flight	Battery

Weather	Temperature	Wind speed	Wind Direction

Crew	
Visiual Observer	Operator
Flight Mission	**Notes**

Drone Flight Log Book

Date:

Drone Model Name

Location (from-to)	Time
Minutes of flight	Battery

Weather	Temperature	Wind speed	Wind Direction

Crew	
Visual Observer	Operator

Flight Mission	Notes

Drone Flight Log Book

Date:

Drone Model Name

Location (from - to)	Time

Minutes of flight	Battery

Weather	Temperature	Wind speed	Wind Direction

Crew	
Visiual Observer	Operator

Flight Mission	Notes

Drone Flight Log Book

Date:

Drone Model Name

Location (from-to)	Time

Minutes of flight	Battery

Weather	Temperature	Wind speed	Wind Direction

Crew	
Visual Observer	Operator

Flight Mission	Notes

Drone Flight Log Book

Date:

Drone Model Name

Location (from - to)	Time
Minutes of flight	Battery

Weather	Temperature	Wind speed	Wind Direction

Crew	
Visiual Observer	Operator

Flight Mission	Notes

Drone Flight Log Book

Date:

Drone Model Name

Location (from-to)	Time
Minutes of flight	**Battery**

Weather	Temperature	Wind speed	Wind Direction

Crew	
Visual Observer	Operator
Flight Mission	**Notes**

Drone Flight Log Book

Date:

Drone Model Name

Location (from - to)	Time

Minutes of flight	Battery

Weather	Temperature	Wind speed	Wind Direction

Crew	
Visiual Observer	Operator
Flight Mission	Notes

Drone Flight Log Book

Date:

Drone Model Name

Location (from-to)	Time
Minutes of flight	Battery

Weather	Temperature	Wind speed	Wind Direction

Crew	
Visual Observer	Operator
Flight Mission	Notes

Drone Flight Log Book

Date:

Drone Model Name

Location (from - to)	Time

Minutes of flight	Battery

Weather	Temperature	Wind speed	Wind Direction

Crew	
Visiual Observer	Operator
Flight Mission	**Notes**

Drone Flight Log Book

Date:

Drone Model Name

Location (from-to)	Time
Minutes of flight	Battery

Weather	Temperature	Wind speed	Wind Direction

Crew	
Visual Observer	Operator

Flight Mission **Notes**

Drone Flight Log Book

Date:

Drone Model Name

Location (from - to)	Time
Minutes of flight	**Battery**

Weather	Temperature	Wind speed	Wind Direction

Crew	
Visiual Observer	Operator

Flight Mission	Notes

Drone Flight Log Book

Date:

Drone Model Name

Location (from-to)	Time
Minutes of flight	Battery

Weather	Temperature	Wind speed	Wind Direction

Crew	
Visual Observer	Operator

Flight Mission	Notes

Drone Flight Log Book

Date:

Drone Model Name

Location (from - to)	Time

Minutes of flight	Battery

Weather	Temperature	Wind speed	Wind Direction

Crew	
Visiual Observer	Operator

Flight Mission	Notes

Drone Flight Log Book

Date:

Drone Model Name

Location (from-to)	Time
Minutes of flight	Battery

Weather	Temperature	Wind speed	Wind Direction

Crew	
Visual Observer	Operator

Flight Mission   Notes

Drone Flight Log Book

Date:

Drone Model Name

Location (from - to)	Time

Minutes of flight	Battery

Weather	Temperature	Wind speed	Wind Direction

Crew	
Visiual Observer	Operator

Flight Mission	Notes

Drone Flight Log Book

Date:

Drone Model Name

Location (from-to)	Time
Minutes of flight	Battery

Weather	Temperature	Wind speed	Wind Direction

Crew	
Visual Observer	Operator

Flight Mission | Notes

Drone Flight Log Book

Date:

Drone Model Name

Location (from - to)	Time

Minutes of flight	Battery

Weather	Temperature	Wind speed	Wind Direction

Crew	
Visiual Observer	Operator

Flight Mission	Notes

Drone Flight Log Book

Date:

Drone Model Name

Location (from-to)	Time
Minutes of flight	Battery

Weather	Temperature	Wind speed	Wind Direction

Crew	
Visual Observer	Operator

Flight Mission	Notes

Drone Flight Log Book

Date:

Drone Model Name

Location (from - to)	Time

Minutes of flight	Battery

Weather	Temperature	Wind speed	Wind Direction

Crew	
Visiual Observer	Operator

Flight Mission	Notes

Drone Flight Log Book

Date:

Drone Model Name

Location (from-to)	Time
Minutes of flight	Battery

Weather	Temperature	Wind speed	Wind Direction

Crew	
Visual Observer	Operator
Flight Mission	Notes

Drone Flight Log Book

Date:

Drone Model Name

Location (from - to)	Time
Minutes of flight	Battery

Weather	Temperature	Wind speed	Wind Direction

Crew

Visiual Observer	Operator

Flight Mission	Notes

Drone Flight Log Book

Date:

Drone Model Name

Location (from-to)	Time
Minutes of flight	Battery

Weather	Temperature	Wind speed	Wind Direction

Crew	
Visual Observer	Operator

Flight Mission   Notes

Drone Flight Log Book

Date:

Drone Model Name

Location (from - to)	Time

Minutes of flight	Battery

Weather	Temperature	Wind speed	Wind Direction

Crew	
Visiual Observer	Operator

Flight Mission	Notes

Drone Flight Log Book

Date:

Drone Model Name

Location (from-to)	Time

Minutes of flight	Battery

Weather	Temperature	Wind speed	Wind Direction

Crew	
Visual Observer	Operator

Flight Mission	Notes

Drone Flight Log Book

Date:

Drone Model Name

Location (from - to)	Time
Minutes of flight	Battery

Weather	Temperature	Wind speed	Wind Direction

Crew

Visiual Observer	Operator

Flight Mission	Notes

Drone Flight Log Book

Date:

Drone Model Name

Location (from-to)	Time

Minutes of flight	Battery

Weather	Temperature	Wind speed	Wind Direction

Crew	
Visual Observer	Operator

Flight Mission	Notes

Drone Flight Log Book

Date:

Drone Model Name

Location (from - to)	Time
Minutes of flight	Battery

Weather	Temperature	Wind speed	Wind Direction

Crew	
Visiual Observer	Operator

Flight Mission	Notes

Drone Flight Log Book

Date:

Drone Model Name

Location (from-to)	Time
Minutes of flight	Battery

Weather	Temperature	Wind speed	Wind Direction

Crew	
Visual Observer	Operator

Flight Mission | Notes

Drone Flight Log Book

Date:

Drone Model Name

Location (from - to)	Time

Minutes of flight	Battery

Weather	Temperature	Wind speed	Wind Direction

Crew	
Visiual Observer	Operator

Flight Mission	Notes

Drone Flight Log Book

Date:

Drone Model Name

Location (from-to)	Time

Minutes of flight	Battery

Weather	Temperature	Wind speed	Wind Direction

Crew	
Visual Observer	Operator

Flight Mission	Notes

Drone Flight Log Book

Date:

Drone Model Name

Location (from - to)	Time

Minutes of flight	Battery

Weather	Temperature	Wind speed	Wind Direction

Crew	
Visiual Observer	Operator

Flight Mission	Notes

Drone Flight Log Book

Date:

Drone Model Name

Location (from-to)	Time
Minutes of flight	Battery

Weather	Temperature	Wind speed	Wind Direction

Crew	
Visual Observer	Operator

Flight Mission   **Notes**

Drone Flight Log Book

Date:

Drone Model Name

Location (from - to)	Time

Minutes of flight	Battery

Weather	Temperature	Wind speed	Wind Direction

Crew	
Visiual Observer	Operator

| Flight Mission | Notes |

Drone Flight Log Book

Date:

Drone Model Name

Location (from-to)	Time
Minutes of flight	Battery

Weather	Temperature	Wind speed	Wind Direction

Crew	
Visual Observer	Operator
Flight Mission	Notes

Drone Flight Log Book

Date:

Drone Model Name

Location (from - to)	Time

Minutes of flight	Battery

Weather	Temperature	Wind speed	Wind Direction

Crew	
Visiual Observer	Operator

Flight Mission	Notes

Drone Flight Log Book

Date:

Drone Model Name

Location (from-to)	Time

Minutes of flight	Battery

Weather	Temperature	Wind speed	Wind Direction

Crew	
Visual Observer	Operator

Flight Mission	Notes

Drone Flight Log Book

Date:

Drone Model Name

Location (from - to)	Time

Minutes of flight	Battery

Weather	Temperature	Wind speed	Wind Direction

Crew	
Visiual Observer	Operator

Flight Mission	Notes

Drone Flight Log Book

Date:

Drone Model Name

Location (from-to)	Time
Minutes of flight	Battery

Weather	Temperature	Wind speed	Wind Direction

Crew	
Visual Observer	Operator
Flight Mission	Notes

Drone Flight Log Book

Date:

Drone Model Name

Location (from - to)	Time

Minutes of flight	Battery

Weather	Temperature	Wind speed	Wind Direction

Crew	
Visiual Observer	Operator

Flight Mission — **Notes**

Drone Flight Log Book

Date:

Drone Model Name

Location (from-to)	Time
Minutes of flight	Battery

Weather	Temperature	Wind speed	Wind Direction

Crew	
Visual Observer	Operator

| Flight Mission | Notes |

Drone Flight Log Book

Date:

Drone Model Name

Location (from - to)	Time

Minutes of flight	Battery

Weather	Temperature	Wind speed	Wind Direction

Crew	
Visiual Observer	Operator
Flight Mission	Notes

Drone Flight Log Book

Date:

Drone Model Name

Location (from-to)	Time

Minutes of flight	Battery

Weather	Temperature	Wind speed	Wind Direction

Crew	
Visual Observer	Operator

Flight Mission	Notes

Drone Flight Log Book

Date:

Drone Model Name

Location (from - to)	Time

Minutes of flight	Battery

Weather	Temperature	Wind speed	Wind Direction

Crew

Visiual Observer	Operator

Flight Mission	Notes

Drone Flight Log Book

Date:

Drone Model Name

Location (from-to)	Time
Minutes of flight	Battery

Weather	Temperature	Wind speed	Wind Direction

Crew	
Visual Observer	Operator
Flight Mission	Notes

Drone Flight Log Book

Date:

Drone Model Name

Location (from - to)	Time

Minutes of flight	Battery

Weather	Temperature	Wind speed	Wind Direction

Crew	
Visiual Observer	Operator

Flight Mission	Notes

Drone Flight Log Book

Date:

Drone Model Name

Location (from-to)	Time
Minutes of flight	Battery

Weather	Temperature	Wind speed	Wind Direction

Crew	
Visual Observer	Operator

Flight Mission | **Notes**

Drone Flight Log Book

Date:	
	Drone Model Name

Location (from - to)	Time
Minutes of flight	Battery

Weather	Temperature	Wind speed	Wind Direction

Crew	
Visiual Observer	Operator
Flight Mission	Notes

Drone Flight Log Book

Date:

Drone Model Name

Location (from-to)	Time
Minutes of flight	Battery

Weather	Temperature	Wind speed	Wind Direction

Crew	
Visual Observer	Operator
Flight Mission	**Notes**

Drone Flight Log Book

Date:

Drone Model Name

Location (from - to)	Time

Minutes of flight	Battery

Weather	Temperature	Wind speed	Wind Direction

Crew	
Visiual Observer	Operator

Flight Mission	Notes

Drone Flight Log Book

Date:

Drone Model Name

Location (from-to)	Time
Minutes of flight	Battery

Weather	Temperature	Wind speed	Wind Direction

Crew	
Visual Observer	Operator
Flight Mission	Notes

Drone Flight Log Book

Date:

Drone Model Name

Location (from - to)	Time

Minutes of flight	Battery

Weather	Temperature	Wind speed	Wind Direction

Crew	
Visiual Observer	Operator

Flight Mission	Notes

Drone Flight Log Book

Date:

Drone Model Name

Location (from-to)	Time
Minutes of flight	Battery

Weather	Temperature	Wind speed	Wind Direction

Crew	
Visual Observer	Operator
Flight Mission	Notes

Drone Flight Log Book

Date:

Drone Model Name

Location (from - to)	Time

Minutes of flight	Battery

Weather	Temperature	Wind speed	Wind Direction

Crew	
Visiual Observer	Operator

Flight Mission	Notes

Drone Flight Log Book

Date:

Drone Model Name

Location (from-to)	Time
Minutes of flight	Battery

Weather	Temperature	Wind speed	Wind Direction

Crew	
Visual Observer	Operator

Flight Mission | **Notes**

Drone Flight Log Book

Date:

Drone Model Name

Location (from - to)	Time

Minutes of flight	Battery

Weather	Temperature	Wind speed	Wind Direction

Crew	
Visiual Observer	Operator

Flight Mission	Notes

Drone Flight Log Book

Date:

Drone Model Name

Location (from-to)	Time
Minutes of flight	Battery

Weather	Temperature	Wind speed	Wind Direction

Crew	
Visual Observer	Operator

Flight Mission	Notes

Drone Flight Log Book

Date:

Drone Model Name

Location (from - to)	Time

Minutes of flight	Battery

Weather	Temperature	Wind speed	Wind Direction

Crew	
Visiual Observer	Operator

Flight Mission	Notes

Drone Flight Log Book

Date:

Drone Model Name

Location (from-to)	Time
Minutes of flight	Battery

Weather	Temperature	Wind speed	Wind Direction

Crew	
Visual Observer	Operator
Flight Mission	Notes

Drone Flight Log Book

Date:

Drone Model Name

Location (from - to)	Time

Minutes of flight	Battery

Weather	Temperature	Wind speed	Wind Direction

Crew	
Visiual Observer	Operator

Flight Mission	Notes

Drone Flight Log Book

Date:

Drone Model Name

Location (from-to)	Time

Minutes of flight	Battery

Weather	Temperature	Wind speed	Wind Direction

Crew	
Visual Observer	Operator

Flight Mission	Notes

Drone Flight Log Book

Date:

Drone Model Name

Location (from - to)		Time	
Minutes of flight		Battery	

Weather	Temperature	Wind speed	Wind Direction

Crew	
Visiual Observer	Operator

Flight Mission   **Notes**

Drone Flight Log Book

Date:

Drone Model Name

Location (from-to)	Time
Minutes of flight	**Battery**

Weather	Temperature	Wind speed	Wind Direction

Crew	
Visual Observer	Operator

Flight Mission	Notes

Drone Flight Log Book

Date:

Drone Model Name

Location (from - to)	Time
Minutes of flight	Battery

Weather	Temperature	Wind speed	Wind Direction

Crew	
Visiual Observer	Operator
Flight Mission	Notes

Drone Flight Log Book

Date:

Drone Model Name

Location (from-to)	Time

Minutes of flight	Battery

Weather	Temperature	Wind speed	Wind Direction

Crew	
Visual Observer	Operator

Flight Mission	Notes

Drone Flight Log Book

Date:

Drone Model Name

Location (from - to)	Time
Minutes of flight	Battery

Weather	Temperature	Wind speed	Wind Direction

Crew	
Visiual Observer	Operator
Flight Mission	Notes

Drone Flight Log Book

Date:

Drone Model Name

Location (from-to)	Time

Minutes of flight	Battery

Weather	Temperature	Wind speed	Wind Direction

Crew	
Visual Observer	Operator
Flight Mission	Notes

Drone Flight Log Book

Date:

Drone Model Name

Location (from - to)	Time

Minutes of flight	Battery

Weather	Temperature	Wind speed	Wind Direction

Crew	
Visiual Observer	Operator

Flight Mission	Notes

Drone Flight Log Book

Date:

Drone Model Name

Location (from-to)	Time

Minutes of flight	Battery

Weather	Temperature	Wind speed	Wind Direction

Crew	
Visual Observer	Operator

Flight Mission	Notes

Drone Flight Log Book

Date:

Drone Model Name

Location (from - to)	Time

Minutes of flight	Battery

Weather	Temperature	Wind speed	Wind Direction

Crew	
Visiual Observer	Operator
Flight Mission	Notes

Drone Flight Log Book

Date:

Drone Model Name

Location (from-to)	Time
Minutes of flight	Battery

Weather	Temperature	Wind speed	Wind Direction

Crew	
Visual Observer	Operator

Flight Mission	Notes

Drone Flight Log Book

Date:

Drone Model Name

Location (from - to)	Time

Minutes of flight	Battery

Weather	Temperature	Wind speed	Wind Direction

Crew	
Visiual Observer	Operator
Flight Mission	Notes

Drone Flight Log Book

Date:

Drone Model Name

Location (from-to)	Time
Minutes of flight	**Battery**

Weather	Temperature	Wind speed	Wind Direction

Crew	
Visual Observer	Operator
Flight Mission	Notes

www.ingramcontent.com/pod-product-compliance
Lightning Source LLC
Chambersburg PA
CBHW071405080526
44587CB00017B/3184